THE SUBMARINE DURING THE COLD WAR

The Submarine During the Cold War

The Men, the Pride, the Threats, and the Disasters

By
Mark Pater Noster

E-BookTime, LLC
Montgomery, Alabama

The Submarine During the Cold War
The Men, the Pride, the Threats, and the Disasters

ISBN: 978-1-59824-892-0

First Edition
Published August 2008
E-BookTime, LLC
6598 Pumpkin Road
Montgomery, AL 36108
www.e-booktime.com

I would like to dedicate this book to all who have inspired me, and to all submariners, past, present, and to those in the future.

May you all have fair winds and a following sea!

Mark Pater Noster

Contents

Acknowledgements

The author wishes to recognize the excellent advice and help provided by his wife Kathleen, my stepson Thomas Kelly, my parents Janete and Caesar Pater Noster and Selene Fields, my sister. Their recommendations were instrumental in the writing of this book. In addition, I could not have even come close to publishing this book without the many hours donated to editing my countless pages by Pam Wolverton.

Without their help, patience and motivation, along with the countless hours of just plain putting up with me this book would not have been possible.

Introduction

The submarine service and its missions have routinely been kept secret from the public. The submarine's mission itself because of its underwater presence only adds to its mystifying existence. The one fact of evidence that cannot be denied is that this service is and will always be a dangerous profession. This book will attempt to de-mystify the submarine service. At the very least it will explain to the reader why during the Cold War the submarine service played such a vital role in keeping the United States safe, but at a cost. Unfortunately, earning and maintaining freedom often comes with a heavy price and the number of lives lost illustrates this point.

A submariner's life in the United States Navy is like no other branch of service. Locked up in a steel tube (sometimes for months at a time) can present a very difficult and unique problem in any emergency. Though a submarine can surface; help in an emergency is still mostly left up to the crew. The submarine, since surrounded by water must have a crew that can act at a moments notice should flooding, fire, or other problems arise. Outside help is usually only an afterthought, and often too late to be of any assistance.

The standard 24-hour day is no longer part of the human existence or way of life. Submarines run on an 18 hour day. The entire day is changed so that there are 6 hours on watch and 12 hours off watch. This makes up a submariner's day. Even the atmosphere is man-made and controlled. Eyes no longer need to see long distances and they quickly abandon the need to do so. Human resistance to colds and sickness goes down because the air is scrubbed clean of all impurities. The overall 18-hour day consists of standing a multitude of various watches (assignments), sleeping, watching movies, studying, and drilling to improve efficiency.

The submarine, because of its missile firing capabilities and stealth, allows it to easily fulfill the requirements to accommodate Theodore Roosevelt's statement, "Speak softly, but carry a big stick." During the Cold War the submarine, its men, missions, and dangers were the main deterrents. To this day the submarine has the best equipped and hardest to reach arsenal of any other branch of service. The ballistic missiles that are carried on the submarines are always on the move and hard to detect, because eighty percent of the time submarines are on patrol. Land based missile systems are easier to detect because they do not move, but the missiles on submarines are constantly in motion because the submarines themselves are in motion. Submarines present the vehicle for them to be fired from and travel in assigned patrolling areas. This gives the submarine a distinct advantage over other missile systems since the submarine presents a constantly moving underwater target. What is needed now and what will be explained in this book are the dangers that should have been prevented.

The submarine service in all its glory must be equipped to have fewer casualties. The submarine service must be more concerned with the hazardous environment its men live in, so that they are not sacrificing their lives to maintain freedom.

The reader needs to understand that much of the submarine service still surrounds itself with top secret clearances, which delays, and makes the information available to the public hard to obtain.

This book will not violate any of the Navy's rules but will still present new information. This information will be both factual research and actual experiences, providing first-hand knowledge in detail about the submarine service during the Cold War. An interesting side note, like my statement above, is the fact that most submariners, regardless of the dangers presented, would not accept any other life than that of being a submarine sailor. It is hoped that the reader will acquire a new understanding of the submarine service, the men who protect their countries in them, and a submariner's life in general.

Chapter 1

The Cold War

The Cold War may not have been a traditional conflict between the United States and Russia, but it was considered to be a major problem for the security of the United States. The United States and Russia themselves were locked in a bitter battle of monitoring each other's military/nuclear capabilities, building submarines and maintaining superpower supremacy. Quite often each side tried to have the upper hand by out performing the other. Quick submarine building was the order of the day, unfortunately this rapid building to stay one step ahead of the other side sometimes cost lives.

The questions to this day are: Did this attitude to build submarines fast, train fast, and keep up with the Soviets cost more human lives than it should have? Did the politics of secret missions at an alarming rate put submariners' lives at stake? It is a known fact that the Cold War dealt with more than the submarine service, but this book will focus mostly on the United States submarine services and the men and technology that made them play such a large role during this era. The submarine service in both countries enjoyed the elite status to which they were assigned, but also shared in the problems and complexities to keep them operational and maintained.

The Cold War began in 1945 after the nuclear bombing of Nagasaki and Hiroshima and ended in 1991 with the collapse of the Soviet Union. Between 1947 and the early 1970's, the submarine service built over 250 submarines to support the Cold War. Russia and the United States built submarines at an unprecedented rate, completing as many as nine a year combined, for a total of over 276 submarines. (*Cold War Submarines* 122) This was far from the norm of less than one every other year that was the standard before the Cold War Era. Technology and new innovations, sonar, noise quieting material, torpedoes, ballistic missiles, nuclear power, and a host of other new ideas were constantly being added. Submarines were at the forefront of this new technology and benefitted greatly from these advances. Some worked and some were discarded as quickly as they were implemented. The submarine transformed from diesel boats that could not stay underwater for any length of time, to new and deadly submarines that could stay submerged for months. They could travel faster, further, and remain undetected longer than any of their predecessors. The diesel submarines of the past had to come to periscope depth (about 35 feet from the surface). They needed to raise a mast called a snorkel to bring in fresh air and exhaust the diesel fumes while they charged their batteries. The new nuclear submarines did not have these limitations because they ran on nuclear power without the need of running a diesel engine. All of these improvements gave the submarine during the Cold War the ability to have a better chance of preventing a war and keeping the peace because they could stay on patrol longer and were much harder to detect.

During the Cold War, Russia had many different classes of submarines. Each class of submarines made changes and improvements, while some carried missiles, others had improved and more advanced technology that included such things as better propulsion systems, sonar, radar tracking devices, and high speed capabilities. Many of the diesel boat-variety was not very reliable as far as submarines go because they were noisy, dangerous and ill equipped to handle emergencies. Some of these classes, the Juliett, Whiskey, Zulu, and Golf classes were older, noisier and easily tracked by the United States Navy.

Many of these vessels were known to have inherent flaws both in the boats and the weapons they carried because the equipment often was not manufactured up to proper specifications. Sometimes faulty welding was discovered and weapons were found to be unstable while being transported. (*Cold War Submarines* 328) These flaws often showed up as fires (poor electrical connections), poor training and construction, movement of weapons not properly secured, faulty equipment and manufacturing failures.

However, in the 1950's, Russia, like the United States, started building nuclear submarines. These submarines of the November, Hotel, and Echo class were quieter, faster, carried long range missiles and were much more of a concern to the United States Navy because they were harder to detect. So much so, that the United States embarked on building perhaps the most potent and famous submarines built during the Cold War, the "Forty-One for Freedom." The United States authorized forty-one Nuclear Ballistic Missile-firing submarines to be built. All forty-one were built and operational between 1957 and 1967. These submarines consisted of three different classes of nuclear submarines; The Washington, Lafayette, and Ethan Allen. These submarines were later overhauled and modified to carry Polaris, Poseidon and finally Trident missiles. Each modification expanded on the distances they could travel giving the United States the capability to reach a target from anywhere the submarine patrolled. For example, the first Polaris Missiles (A1) had a range of 1,200 nautical miles, while the second revision (A2) had a range of 1,500 nautical miles, and the third revision (A3) had 2,500 nautical miles. This was quickly followed by the Poseidon missile with a range of 2,500 nautical miles, and was able to carry triple the amount of warheads that the Polaris missiles carried. It was perhaps the most ambitious undertaking of any country and because of this became the backbone of the United States Ballistic Missile program. Their firing capability was the major deterrent the United States would have against the Soviets.

Russia refused to slow down; instead, it added its own answers to the American advances by building several newer classes of submarines. These newer subs, the Yankee, Delta, Foxtrot, and Victor III classes were counterparts to America's expanding fleet, and equally potent, carrying missiles that matched or were the equivalent to those on the United States submarines. The United States, finding itself on equal terms with Russia, continued building with its next major class of submarine called the Ohio class in 1979. This new class of submarine was larger, faster, and carried the next generation of missiles called Trident with a range of over 2,500-3,500 miles. (*Cold War Submarines* 190)

The advent of the world's first nuclear power submarine in 1954 opened up a new era in underwater warfare and also opened up a race between the United States and Russia for submarine supremacy. Admiral Gorshkov, Russia's submarine advisor, continued to deploy his submarines regardless of condition in an attempt to keep pace with the United States. It did not matter whether it was the loss of a submarine or human life; he continued to build submarines as fast as possible. (*Cold War Submarines*) Submarines such as the Russian K-429, K-219, K-278 (Komsomolets), K-19 of the Yankee, Charlie 1, Victor, and Hotel classes and many others would bear this out with fires, sinking and loss of life. (*Cold War Submarines* 144) Lack of funding and proper maintenance took second place when it came to getting a submarine to sea, which was the only issue as far as the Soviet Navy was concerned.

Directors of the National Security Agency, William Burr and Thomas Blanton, reported in "U.S. and Soviet Naval Encounters during the Cuban Missile Crisis," that many of the Soviet submarines patrolling near Cuba had nuclear-armed torpedoes. A major confrontation between the United States and Russia during the Cold War years occurred during the Cuban Missile Crisis as was reported in the National Security Archive Electronic Briefing Book from October 1962. The authors reported, after researching declassified material, that one of the incidents involved a Russian submarine, the B-59 a Foxtrot class, which had been forced to surface on October 27, 1962 just after the Soviet's shot down a U-2 over Cuba. This sub came extremely close to attacking a U.S. Naval vessel with a nuclear weapon. The United States was not aware that Russia had so many submarines with nuclear weapons aboard patrolling in and around Cuban waters. In fact, they were not reported until three Soviet submarines were ultimately forced to the surface by the United States during the Crisis. The submarines were surrounded by various components of the United States Naval Fleet and were constantly harassed to the point that Russia decided it was far safer to surface than stay submerged and possibly come under attack by United States Forces. (*Cold War Submarines* 204) Khrushchev believed that it was proper for his submarines to sink the ships of the United States if they were stopped or blockaded. (*Cold War Submarines* 204) The report goes on to state that "No one on the U.S. side knew at that time that the Soviet submarines were so physically difficult and unstable that commanding officers fearing they were under attack by the U.S. forces may have briefly considered arming their nuclear torpedoes." (*Cold War Submarines* 204) The Russian submarines were under so much pressure that they were unsure of just what the United States intended to do. Many of the Russian skippers felt pressured to act regardless, and were told not to allow the blockade to stand.

During the month of October 1962, while the intensity between Russia, Cuba, and the United States was at a peak, new and critical information was obtained which confirmed that many Russian subs were armed with nuclear weapons. Just how close the United States may have been to a nuclear confrontation will never be known, but it was much closer than anyone would have thought or knew about. Until the United States became proactive and requested Cuba and Russia to withdraw the missiles just how many Russian submarines were actually involved was quite a surprise to the United States.

The two most powerful countries in the world played out the Cold War. Both countries were constantly gathering and updating intelligence on each other. The amount of firepower both countries had during the Cold War was enough to destroy the world as it was known. The fact these weapons were kept in check is evidence by itself of the mutual respect and possible fear each country had for the other. (Polmar, Norman and Moore, Kenneth) During the Cold War in 1962 the Cuban Missile Crisis was definitely a major concern faced by both countries. Historian John Lewis Gaddis states, (*The Cold War:*

A New History) "What kept the war from breaking out in the fall of 1962 was really irrationality, on both sides, of sheer terror." He also states that the "Cuban Missile Crisis only confirmed how difficult the task would be." One lesson that came out of it was the extent to which the Russians and Americans had failed to think rationally going into it.

While the Cold War was acknowledged by the general public, and often spoken about in the news, the public did not understand the seriousness that it presented nor how close to a nuclear war Russia and the United States had come. Many Cold War historians such as Bently (*The Death of the Thresher*) and Bowermaster (*The Last Front of the Cold War*) were equally left to ponder the overall results since so little was ever disclosed openly because of its classified and secret nature. Many historians are only now beginning to be able to obtain submarine reports that have become declassified to the public such as the reports and hearings about the USS Thresher released from the Joint Committee on Atomic Energy.

Historians on the Cold War, such as John Gaddis, Lawrence Freedman and Vladislav Zubok continue to discuss why the Cold War finally came to an end. It obviously was not new knowledge that Russia's economy was in ruins towards the end of the 1980's. Whether this played a role in the collapse of Russia and caused better relations with the United States is open to speculation. The fact remains that in 1991 the Soviet Union did collapse and the economy was at an all time low, obviously playing into part of the reasons for the collapse.

Submarine building and maintenance costs, and Russia's lack of money and support for the submarine service, made it hard to keep the submarine program funded and operational. This also contributed to the demise of the Soviet Union, and certainly led to the terrible shape that the submarine service found itself in at the end of the Cold War in 1991.

Chapter 2

The Submarine Deterrent

The submarine easily adapts to its role as a deterrent and no other military ship spreads as much fear as a submarine. The stealth and underwater habitat that surrounds the submarine service today make it an unusual weapon. So much so that it is the perfect weapon to surprise and keep enemies from carrying out an act of war. Virtually untraceable and unknown as to its whereabouts, it spreads instant fear into any adversary. The mere fact alone that a submarine can bring more firepower than all the bombs dropped in the Second World War combined makes one think twice before contemplating an act of aggression. Many of today's weapons have this capability, but the small submarine is considered one of the most potent. (Undersea Stealth 164) Russia built submarines in the 1960's and 70's specifically to counter the perceived threat from the United States, and the United States built submarines to ensure Russia was kept at bay. Two major superpowers kept each other on equal ground defending their own beliefs. The threat was pretty much neutralized because both countries respected each others submarine capabilities.

Theodore Roosevelt, many years ago, felt that war could be prevented if you exhibited yourself as a strong force. Historian, Thomas Bailey, who disagreed with Roosevelt's policies, nevertheless concluded that, "Roosevelt was a great personality, a great activist, a great preacher of the moralities, a great controversialist, and a great showman. He dominated his era as he dominated conversations." As long as the United States had a strong hand he felt that there was no better position to keep the peace and that war could be prevented. Roosevelt's strong stick policy often heard in his speeches did not advocate war but outlined his beliefs that preventing one required having a strong military presence. President Roosevelt had plenty of experience as the Assistant Secretary of the Navy in 1898, followed by the Vice presidency until he became President in 1901 taking over after the assassination of President McKinley. If the balance of power ever changed, it could have meant the downfall of Democracy in the United States; fortunately this country never had to find out.

Submarine patrols were almost certainly constant, but just how many were actually completed by both countries may never be known, although the U.S. Navy has now declassified and released 23 years of patrol data for Soviet/Russian submarines covering 1956 through 1979. A follow-up request by the author for what constituted a "patrol" triggered the following response as stated by The United States Navy's Office of Naval Intelligence (ONI) "We cannot release specific criteria for determining what a 'patrol' is as it would divulge methods and sources." Instead they released to the public the number of patrols completed to defend this deterrent force, but not what a patrol actually consisted of or where patrol areas were located. This, as it is now, was classified information, though some reports are finally being released. The total number of patrols during the Cold War between both countries is still open for speculation, though the United States has released the total completed by its submarines.

The ONI also pointed out that Russia completed approximately 230 patrols during 1984. The ONI further stated how much the Russian submarine force has deteriorated since the Cold War ended. As stated earlier equipment failures and lack of proper maintenance due to funding constraints caused much deterioration in the Soviet Fleet. This caused the deterrent status of Soviet/Russian submarines to also deteriorate and compromised the once most powerful submarine service in the world.

The United States also started to scale back on its operations as well in the late 1980's, realizing that the Russian threat was no longer even close to what it once had been. The United States submarines between 1947 to 1991 made over 3,500 successful strategic deterrent patrols and uncounted surveillance and barrier patrols without major equipment failures or problems. They were also responsible during these major campaigns, such as Korea and Vietnam, for making many offensive, defensive, and special operations patrols. Russia, according to the ONI, came close to the same number of patrols (3,400) during this time period, but had many patrols that had to be shortened due to equipment problems.

The United States prepared for a war that never came, and the submarine force was able to take the special trust and confidence in the abilities given to them and helped to maintain the peace, and while there were many conflicts, they were never waged on American soil. The United States went to sea entrusted with weapons of awesome destructive power, and with the sole purpose to prevent war. This was the so-called mission of our deterrent force. For the entire Cold War years it was obviously very successful. The Ballistic Missile submarine was a major and survivable leg of this country's strategic triad. Other locations for this triad were known, but the submarines were always on the move, thus making it a harder target to detect or reach. The United States had the upper hand in this war of nerves because our submarines were in better condition, better maintained, and had far less problems than our Russian counterparts.

The first nuclear Ballistic Missile submarine "The George Washington" proved that the nuclear age as a deterrent to war had arrived, when she successfully fired the world's first ballistic missile from a submerged platform on June 28, 1960. This was the beginning of the largest submarine building-era between the two largest super powers in the world. Though, some minor concerns and arguments materialized through the years, both sides ultimately were successful in maintaining this balance of power. The George Washington left Charleston, South Carolina, in 1960 with our first deterrent patrol and what was to be the first of thousands to follow. The Washington was the first of a five-ship Washington class of submarines (often called Boomers), which the shipbuilders at Electric Boat Company altered from the design of an attack submarine then under construction (USS Scorpion SSN 589) and added over 125 feet for the missile compartment.

When completed, the USS George Washington (SSBN 598) measured 382 feet long and 33 feet abeam, displacing 6,709 tons of water submerged. Her surface speed was 16 knots, substantially slower than her submerged speed of 22 knots. (*Submarines War Beneath the Waves* 170, 171) She was faster submerged because she presented less friction while underwater. Commissioned at the end of 1959 the George Washington went to sea on her first patrol in November 1960 with a crew of 112 and 16 Polaris A-1 ballistic missiles. The submarine was refitted with improved missiles several times and continued patrols until the early 1980s. In her latter years she was used as an SSN (fast attack submarine) no longer using her missile firing capabilities. She was decommissioned in 1985 after some 26 years of service.

George Washington SSBN 598 at Launching in 1959
(Picture courtesy of U.S. Navy)

Forty more submarines were added to the United States Ballistic Missile submarine force over the next several years. This force was aptly named the "Forty-One for Freedom" as they made up the majority of our deterrent force against any adversary. These submarines consisted of five total classes, which will be outlined and explained, in the creative portion of this book. All five classes were fitted with 16 ballistic missiles and were home ported so that they would be able to hit targets that would cover any target the United States deemed necessary. All of these submarines exceeded their life expectancy which was 20 years. In the latter years, instead of going straight into retirement some of the submarines were used to test new systems and assigned new jobs to keep up with the changing times. These included being converted to SSN's (fast attacks) and special force carriers for the Seabees and Navy Seals. This flexibility further enhanced the United States' deterrent force by providing new methods and uses for these submarines to keep up with the forever-changing world. The added roles allowed the United States to keep these submarines in service longer without compromising their mission.

Jon Bowermaster, a researcher and historian for the Atlantic Monthly, wrote an article on foreign affairs called "The Last Front of the Cold War," in the November 1993 edition. Bowermaster points out how often the United States trailed Russian submarines on their patrols. At times this became a very dangerous mission for both sides. According to Bowermaster on the afternoon of March 20, 1993, a pair of submarines collided seventy-four meters (242 feet) beneath the icy waters of the Arctic Ocean. The concern was that even though these two opposing submarines were not lost, it could have been much worse. What Bowermaster described was the "fact that the USS Grayling, the United States (fast attack) submarine, actually struck the Russian submarine while engaged in surveillance." He further stated that "had the U.S. submarine been just five seconds slower it would have hit the Russian submarine in its missile bay, conceivably opening a crack where the missiles were stored, sinking the sub and scattering nuclear warheads over the ocean floor." (*The Last Front of the Cold War*)

Though this incident ended without further serious repercussions, what could have been is extremely alarming. The damage this accident could have caused and what the radiation levels might have been has always been open to speculation. A good question to ask here might be how this could even happen, given the state of high technology and advances in radar and sonar abilities that both countries had at their disposal? More importantly is the fact that human life once again could have been lost. The United States and Russia played a game of trailing one another and taking pictures of each other while in very close proximity and very often put themselves in danger by doing so.

What is known is that "MAD" (Mutually Assured Destruction) probably played a big part in ending the Cold War. No country will ever be completely safe as long as there are nuclear warheads in the world so it is important to maintain a deterrent status. Since the Cold War ended there has been a scaled down fleet in the United States for Ballistic Missile Submarines. Today, the United States has 19, which is down from the 1960's and 70's, but it still has many of the same concerns, except that now Russia is no longer the same or only threat since other countries are now starting to develop formidable submarine services of their own. It is imperative because of these new challenges that the United States not let down its guard and maintains a strong deterrent force like it did during the entire Cold War Era.

The United States of America and the Soviet Union on May 26, 1972 took steps to limit the amount of Anti-Ballistic missiles (ABM'S). The President of the United States Richard Nixon and the General Secretary of the Communist Party of the Soviet Union Leonid Brezhnev signed an Anti-Ballistic Missile Treaty. This treaty was considered by the United States as a way to maintain an even balance of power between the two countries and to limit any nationwide ABM defense base from being established. The fact that the Cold War was over since 1991, and both countries had already scaled back the number of missiles, really made this a non-issue. What it did however, was to limit the number of nuclear weapons for each country during the remaining Cold War years from 1972-1991. There were various concerns on just how effective this treaty was, and how well it was followed by both sides. Also, there was the added concern that new missile systems under construction would circumvent the treaty even though one of the purposes of the treaty was to prevent this from happening.

The fact remains that nuclear warheads and nuclear submarines have an increased concern when the safety of the crews are involved. The deterrent aspect presented in this paper relied upon both Russia and the United States adequately monitoring and maintaining a visual watch over what the other one was doing. During the Cold War it is safe to say that both countries worked hard to maintain the deterrent force that each side felt was needed. This awesome power that both countries had kept this deterrent force from ever having to be put into action.

Chapter 3

The Nuclear Years

Admiral Hyman G. Rickover, often called the "Father of the United States Submarine Nuclear Navy," had a vision and a plan that ultimately paved the path for the nuclear submarine to get under way. Once he made his recommendations (submarines need to run on nuclear power) he fought to make this happen and refused to take no for an answer. The submarine service changes that took place between 1954 and the mid 1970's would change the way submarines would operate. Rickover was the drive behind the nuclear submarine Navy and the changes that would take place were because of his refusal to except mediocrity.

The changes that took place not only revised the way the submarine was used but also where they were deployed. The limitations of fuel and water storage prior to the nuclear age (they had to be stored) were no longer applicable because the submarine no longer relied on carrying fuel and water. The submarine could make its own water, purify its own air and travel on its own power without the need to surface, run a diesel to charge batteries, or stay close to shore for supplies.

Nuclear power, therefore, transformed the diesel submarine into a submarine that could stay underwater, undetected for months at a time. Food and man himself were the only major concerns as to how long a submarine could remain at sea.

Overseas Naval bases were added in coastal ports of friendly countries to support these submarines and to allow for the capability of striking and reaching anywhere in the world should the need arise. During the building of our ballistic missile submarines, bases and homeports were added (during the 1950's and 1960's) overseas to support these submarines. These bases however, were not always met with the enthusiasm that the United States had hoped for. Andrene Messersmith, a young girl that grew up in the shadows of the Navy's new base that opened in Holy Loch, Scotland explained her views in a book entitled *The American Years*. She described in great depth the concern that developed when the United States decided to forward deploy several submarines with the base opening of Holy Loch in Scotland in the early 1960's.

Though this would change the little town of Dunoon (Population 8,251) (2001 Census) for the next thirty years it did not start off without having much skepticism and tension. Protests in and around Holy Loch, Scotland and also in Rota, Spain, both new submarine bases, in the early years often disrupted the movement of submarines. Protesters went as far as blocking entranceways and harassing the submarine sailors. They protested by holding up signs against the submarine base in large crowds whenever our submarine sailors went ashore. The protests in Scotland were against nuclear power and included any methods that could be used to disrupt the movement of the United States submarines in and out of the Loch, and in several incidents sailboats were used to block the path of the United States submarines. The townspeople were not thrilled about having a nuclear submarine base or nuclear weapons "on their doorstep." (Readers Digest July 1963) In addition, the competition of Americans dating local women in the small towns in and around Dunoon was not taken lightly and often was the gossip of the local newspapers or at town meetings. (*The American Years*) The American presence was well recognized for what it meant to the local economies and the locals found themselves in the unhappy position of wanting the money and extra business from the American presence, but not liking what it represented in their towns. Needless to say the Navy had a huge influence on the little town of Dunoon, and there wasn't a time of day where the Navy's presence in the town and shops was not felt. This arrangement would continue for thirty years and both parties adjusted accordingly.

The fact that nuclear power allowed benefits and comforts not previously enjoyed by diesel-electric submarine counterparts made living on nuclear submarines much more enjoyable. Fresh water was actually made from sea water and the water usage problems of the old World War II subs were no longer an issue. Oxygen was actually removed from the sea water in a process called electrolysis, where the oxygen and hydrogen were removed and stored. The hydrogen eventually would be sent back into the ocean, and the oxygen was stored in large tanks (banks) for breathing. In addition the air would be scrubbed clean of carbon dioxide and purified. Atmospheric monitoring would ensure that the air was safe and breathable for all the men aboard.

Propulsion was also changed, instead of large diesel engines requiring air and oxygen to run, nuclear power ultimately became the power responsible for turning the boat's propellers. Nuclear power safely created steam that would drive large steam turbines. These turbines, through a series of reduction gears and other propulsion equipment, would output to the propeller or propellers. Through the use of two separate loops and adequate shielding the propulsion plant was a marvel of technology and extremely safe. In fact, the system was so much safer that the average radiation level humans received while operating the submarine was less than what would be received from the sun.

Living conditions also were much improved with regular sleeping quarters, air conditioning, stereos, movies and food cooked the way the crew liked it. Pizza, steak, and lobster nights were favorite pastimes and were rarely missed. While many pastimes to keep busy when not on watch, like reminiscing and joke night competitions, were carried over from past practices, new ones were often invented to fit the situations. Halfway nights (a party thrown when the patrol reached the halfway point of the cruise) became special events that helped to pass the time, and of course the long-standing tradition of playing cards is not to be forgotten. The missile room, which was about 100 feet long, often became the running track and exercise room. Many of these pastimes were limited in the diesel subs because of their small size, but now were utilized more often because of the extra room that these newer submarines had to offer for human comfort.

Watching, listening and maintaining the stealth and unique atmosphere of the submarine was, however, where the most time was spent. Noise was always kept to a minimum, because noise travels through water to great distances, and could give the submarine's position away to an enemy that was miles away. Finally, many different types of drills that included practice for flooding, fire, and smoke were an everyday occurrence, and would keep the crew alert and in top efficiency. A submarine crew has to be ready at a moment's notice to fight any type of hazard whether it is fire, flooding, gas, chemical, or radiation. Constant training, with at least one drill a day, prepared every crewmember and ensured success if such an emergency arose. In addition several hours of study on the ships systems were part of every crew member's daily routine.

The type of nuclear power plant under consideration was the first obstacle to over come. Otto Hahn and Fritz Strassman first discovered nuclear fission in 1938. Many experiments conducted by nuclear physicists and naval personnel in labs and on military bases were used to determine how best to harness the heat source and convert it into propulsion safely. Other studies were also being conducted at the Kaiser Wilhelm Institute, Colombia University and at the Naval Research Laboratory. (*Cold War Submarines* 49, 50) Thermal fission and fast fission using enriched Uranium were the prime study areas and were considered the most successful methods. Transferring heat from the reactor core to a propulsion plant needed to be improved, and both were considered for the process. Under consideration would be pressurized water and pressurized gases. In the end pressurized water was the most attractive. By using a steam generator that isolated the two individual loops from the other the submarine reactor could now be used for turning a propeller. The diagram below, from the Nelson Institute of Marine Research by Stephanie Kellerman, demonstrates how a submarine reactor is used to turn a submarine's propeller.

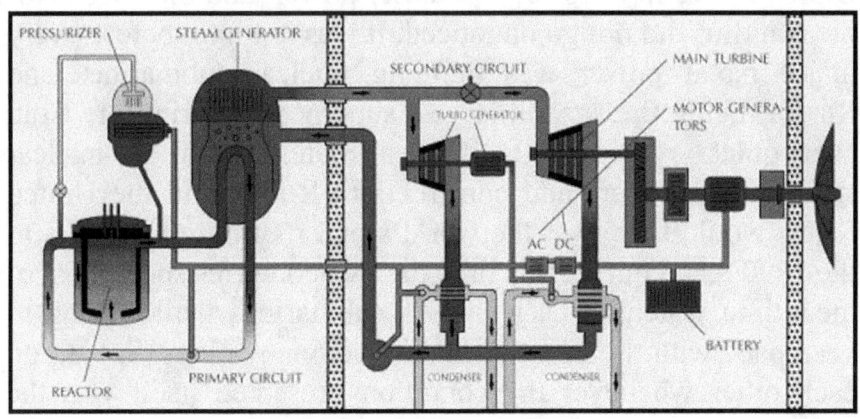

The primary closed loop is the contaminated loop. This loop consists of high-pressure coolant heated by the reactor core, pumped through the steam generators and returned by coolant pumps. The secondary is a low pressure loop that also passes through the steam generator but is separated from the primary loop. This loop feels the effect of the heat made in the primary loop and changes the coolant into steam to drive the steam turbines that drive the propeller. All that is left is to change this steam back to a liquid state. This is done by passing the steam back through large condensers that are open to the effects of cold seawater. Both loops work independently from each other and both are constantly re-circulating, the primary, high pressure coolant and the secondary steam. This method of using a pressurized thermal reactor proved to be an efficient and highly reliable and controllable method for nuclear propulsion and is still in use today. (*US Nuclear Submarines: The Fast Attack*) (6-8)

Nuclear propulsion all started with the commissioning of the USS Nautilus back in 1954, (the world's first fast attack submarine) and in 1959 the USS George Washington became the first Nuclear Ballistic Missile submarine. No longer did the

United States have to rely on diesel power, and the success of the Nautilus did not go unnoticed. It was not long before every major naval power was building nuclear submarines and Russia, being the second largest superpower during the Cold War quickly became the United States' equal in nuclear submarine numbers and construction. Russia and the United States would take over the world's oceans from the 1960's to the 1990's and during this time the two countries spent most of their time watching each other's submarines while trying to keep pace with the latest technological innovations. Spying on each other whenever the condition presented itself was the order of the day, and often, a high priority for both sides. The factual information to support this statement is classified, but is widely known by anyone who served on submarines. Russia built many classes of nuclear submarines during the 30 years of the Cold War. While many were not very reliable as has been previously noted, Russia continued to experiment with noise quieting hulls and different propulsion plants. Large double hulls and sometimes two nuclear reactors were often fitted on the Russian boats. The Russian Sierra Class submarine, a new class of submarine, even used titanium hulls which made this class capable of diving to unprecedented depths. The titanium hulls had more flexibility and were better suited for submarines. Russia's Akula I and the revised Akula II class of nuclear submarines along with the Oscar class (also new to the Russian submarine fleet), were considered the mainstay of the Russian threat during the Cold War. These three classes were often spotted by our own submarines during various exercises and were often the cause of extended submarine missions.

Nuclear missiles and nuclear tipped torpedoes also played a large role in the deterrent forces for both countries. It was quite evident to both sides that with all this fire power (more fire power than all the bombs dropped during World War II) the release of these weapons against each other would mean a quick and final destruction with no winners, as both sides would be destroyed. Since neither country had the upper hand it caused both countries a moment of pause before any ideas of using them would ever have been established. There would be no advantage for either country to use these weapons if it caused the destruction of both countries and both countries had the full capabilities of being able to strike each other from any area these submarines patrolled. The submarine population grew during the 1960's and continued to grow well into the 1980's.

Unfortunately along with this growth governments experienced several submarine accidents and causalities. Some would cause damage to the submarine; others hurt or killed the men who served in them. Many were known and talked about, others have been kept secret. The one truth that is acknowledged is if more submarines are in the oceans there are better chances of submarine disasters. Nuclear power changed the submarine community. It brought about a true submarine that could now stay underwater longer and travel faster than ever before. Crew comfort was drastically improved as well. Propulsion also saw drastic changes as the diesel engines now became secondary power sources next to the nuclear reactors. Along with these changes nuclear power also saw protests and a concern for the risks that came with it. Yet, the countries that switched over to nuclear power soon realized that the efficiency and reliability far outweighed any disadvantages that came from its use and quickly embraced this new source of power for the submarines of the future.

Chapter 4

The Thresher Disaster
A Case Study

On April 10, 1963 at 9:14 in the morning the USS Thresher reported that it was proceeding to test depth. (The deepest depth that a submarine can go to safely according to its building plans) At 9:17 the Thresher reported on its underwater telephone that she was having minor problems and was attempting to blow its ballast tanks. The next sound heard was the unmistakable sound of a submarine breaking apart. The Thresher had sunk in 14,000 feet of water, far below its crush depth. (The depth a submarine implodes.) Somewhere between 9:17 and 9:25 on that horrible morning the Thresher and her crew of one hundred and twenty nine completed their last dive and would never be heard from again.

The Thresher SSN-593 has been the source of much criticism and speculation throughout the years as to what really happened to her. Regardless, she is now at the bottom of the ocean. Many arguments have been developed which state equipment failure was the culprit. Reading through the published research has not made it any easier to determine the real cause of her demise. Critics say the answers are known but will not be released to the public. Piecing together the events from the beginning will shed some light on this subject. The official Navy reports from the JCOAE (Joint Committee on Atomic Energy) left a lot to be desired. The biggest concern that it left was the fact that improper or inadequate ship building and testing caused the Thresher's loss.

The official report from the Navy and the Joint Committee on Atomic Energy lists the loss of the Thresher as unknown, but it also states that the most likely cause was failed silver brazing and welding of a pipe joint. This caused a domino effect by high pressure water spraying on electrical equipment that caused it to short out. The safe guards built into the reactor control caused the reactor to scram (shut down). This caused the submarine to loose propulsion, which it needed to drive the submarine to the surface. In addition it was determined by this report and voice contact with the Thresher, before the loss, that the emergency blow system (high-pressure air blown into the ballast tanks to quickly achieve positive buoyancy) failed to operate properly. The report found that the area in and around the reactor compartment was built to a higher standard than the rest of the submarine. In addition it also found that there were known problems in the way welded joints were tested.

The Thresher, built at the Portsmouth Naval Shipyard in Kittery, Maine was the newest, fastest, and deepest diving submarine built to date for the United States Submarine Service. She had the most advanced technology and the latest weapons system. She was clearly the most up to date Fast Attack Submarine ever built in the United States. Historian Norman Polmar's book *The Death of the USS Thresher* stated "In the United States Navy, the "ultimate" hunter-killer submarine was the Thresher." Sixty years of submarine technology, along with the experience of two world wars, and the knowledge obtained by leading the world in nuclear power development went into the Thresher design according to Polmar, who was a consultant to the Navy and Department of Defense and was a member of the Secretary of the Navy's Research Advisory Committee (NRAC).

The Thresher started her life with the first sections of her keel laying on May 28, 1958, at Portsmouth Naval Shipyard in Kittery, Maine. She would measure 278 feet, 6 inches long. Her circular hull at the widest point would be 31 feet, 8 inches. A little more than two years later, on July 9, 1960, Thresher would be launched with only a year to go before she would be commissioned. She took her name after the "Thresher" which is a long-tailed shark.

On April 30, 1961, Commander Axene, the Thresher's first captain, took his ship to sea for her initial cruise. Commander Axene had watched the Thresher being built from scratch and was satisfied with her progress. The commander knew the sub well and was ready to put her to the test. Thresher made her first dive in relatively shallow water over the continental shelf to test the submarine's sea worthiness, trim, and control. When this initial dive concluded without problems Commander Axene decided to take the sub for its first deep dive test the next day, which was a normal practice. Unfortunately, this was the beginning of several major problems for the submarine. During this test the submarines diving depth gauges read incorrectly and required the submarine to halt the dive and come back home for repair and to re-calibrate its meters.

Commander Axene took no chances with his $45 million submarine. Once back in the shipyard the gauges were modified and Thresher returned to sea two weeks later and had no further problems. The Thresher was now ready to begin her service life and on August 3, 1961, was commissioned into the active fleet of the United States Submarine Service. At this point the submarine was determined to be ready for shakedown cruises which would insure all of the submarine's equipment performed to proper specifications. The Thresher's first assignment was a test of the submarine and its crew. It permitted the submarine and crew to test the capabilities they were designed for, and also allowed the crew to train and prove the compatibility between man and machine. Thresher was built to be extremely quiet, and the shakedown cruise showed evidence that she was even quieter than expected. After this cruise in September 1961 the submarine returned to Portsmouth for a limited maintenance period.

In late October 1961, after Thresher's first limited maintenance check, she once again put back to sea and pulled into San Juan, Puerto Rico traveling some 1,700 nautical miles. It was here that she experienced another major problem. Submarines shut down their reactors while in port and usually rely on shore power. This is where electricity is supplied to the submarine through the shore facility. San Juan did not have the capability to provide shore power so the submarine decided to run its diesel generator while in port to provide power while the reactor was down. About eight hours after the reactor was shut down the diesel engine broke down. This caused the submarine to run off its batteries until another submarine was able to tie up alongside and help by running its diesel engine to provide power.

During the time the Thresher was on its batteries, inside temperatures became extremely high; in some engineering spaces they exceeded 120 degrees. It was not until the submarine could get the reactor back up that things once again returned to normal. Though the port was not able to provide electrical power to the submarine the diesel engine should still have been able to run without incident. It should not have caused the crew on the submarine to work most of the night and well into the next day until the diesel engine was finally repaired. (*The Death of the USS Thresher*) Over the next several months the Thresher would continue to undergo many other tests and modifications necessary to improve the shock mountings on some equipment. The shock mountings provide equipment with shock absorber effect and noise isolation from the rest of the ship. Most shock mountings are made of hard rubber between the mounting plates and equipment. These improvements meant the equipment was now able to withstand any shock forces that could happen during an explosion.

Many tests were conducted on the submarine to test noise output both before and after these modifications were installed. Thresher, during this testing, was subjected to a greater amount of pounding (explosions near the submarine) than any other submarine in the history of the United States Navy. The Thresher continued training and testing for most of 1962, and at the end of the year she was back in Portsmouth Naval Shipyard for overhaul. It was noted that though shipyard relations was and always had been excellent, during the latter part of this overhaul shipyard relations had deteriorated. The crew was especially not happy with the way the shipyard employees left the submarine filthy after completing work. This led Commander Axene to make the following statement, "It is true that we felt they should have been more efficient, should have done better work at times, and should have done a better job of cleaning up after themselves." (*The Death of the USS Thresher*)

Many holes were cut into the submarine's hull during this overhaul to allow for equipment removal and replacement. Whenever this was done the hull had to be inspected and xrayed once it was welded back together. In April 1963, the submarine overhaul was finally reaching its completion after nine months and Commander Axene was ordered to a new command. The Commander later stated he was not happy with these orders, as he preferred to stay with the Thresher. (*The Death of the USS Thresher*) The new captain of the Thresher would be Lieutenant Commander John Wesley Harvey who was a Naval Academy graduate and thirty-six years old.

During the nine-month overhaul, Thresher's crew submitted 875 work requests to be completed (These are problems or requested modifications to equipment.), and all but five were completed satisfactorily. When a work request was completed it had to be approved by both submarine personnel and shipyard supervisors. The five requests not completed were considered minor and not important enough to delay the upcoming sea trials that marked the end of the overhaul.

On April 9, 1963, at approximately 3:45 AM the submarine was once again making preparations to go to sea. At 6:15 that morning, the Executive Officer reported to the Commanding Officer that the reactor was critical. This meant the reactor was self-sustaining, which meant the reactor could maintain nuclear fission. This would allow the subs turbo-electric generator to produce the electricity needed for every piece of equipment on the submarine. At 7:30 the engineering department reported that the submarine's propulsion plant was ready to answer orders, meaning Thresher was ready to get under way under her own power. The submarine slowly pulled away from Pier 11 at the shipyard and headed out to sea away from the New England coastline. Later that morning she would rendezvous with the submarine rescue ship Skylark.

Skylark would maintain communications with Thresher and provide whatever assistance she could to the submarine.

The Skylark was designed to help men escape from a submarine if necessary and was capable of hovering over a sunken submarine. She also had an escape diving bell that could be lowered over an escape hatch of a downed submarine to rescue submariners trapped inside. After an initial shallow dive without any problems the Thresher came to the surface and released the Skylark, requesting her assistance again the next morning when she would attempt her first deep dive since the overhaul. That same afternoon Thresher quietly slipped below the surface and headed for the deep diving test site that took her past the 600 feet limitations of the continental shelf. In route she conducted some high-speed runs (greater than 20 knots) and calibrated equipment in preparation for her test dive the next morning.

At 6:35 on the morning of April 10, 1963 the submarine came up to periscope depth some ten miles away from the Skylark, that was again in position to offer assistance if needed. A submarine comes up to periscope depth so she can raise her antenna masts and periscopes above the water level at the ocean's surface, normally about 28-32 feet. This would be the last time the Thresher would ever see periscope depth as she made final preparation to dive to her test depth.

Test depth is the deepest depth that a submarine is able to go safely without endangering the submarine or its crew. Below test depth, a submarine's piping and systems start to break apart, risking the submarine going to crush depth, where the hull implodes, killing all instantly. The particular numbers of how fast and deep a submarine can dive is classified, but is thought to be below 400 feet and faster than 20 knots.

At 7:47 Thresher, using an underwater telephone, notified Skylark that she was preparing to dive to her test depth. At 7:53 the submarine reported she was leveled off at 400 feet and was checking for leaks. When a submarine dives it usually goes down in increments, levels off and checks all compartments one at a time for normal conditions on the dive. It does this again for each new depth that it attempts to reach, and unless the submarine is in a fight, no depth changes take place without the complete 100% check of the entire submarine. Proper communications and reports must also be given. At 7:56 the Thresher communicated with the Skylark and informed her that all future depth reports would be in reference to test depth. This was done so as not to give away her classified test depth should anyone else intercept her communications. At 8:09 Thresher reported she was at half her test depth. At 8:35 she again reported, stating she was -300 from test depth. According to the Skylark logs at 9:02 the Thresher communicated with her, asking for a repetition of a course. Approximately two minutes later the Thresher ran into serious trouble. According to the Skylark and Commander Hecker, Skylark's captain, the next several minutes were an assortment of unclear and confusing communications. At approximately (some discrepancies in the logs) 9:15 the Skylark stated that she believed to have heard a communications sent from the Thresher that said the submarine was attempting to emergency blow and force itself to the surface and obtain a positive up angle. What this means is that the submarine was attempting to use high pressure air blown into its ballast tanks in an attempt to get the submarine to rise, usually in a hurry, and usually for emergencies. A positive up angle means the sub is on the rise with the bow at an up angle in relationship to the stern. Several of the officers on Skylark reported they heard slightly different reports, but all reported the fact that the submarine was having trouble at presumably test depth and attempting to blow ballast.

Understanding how a submarine dives and surfaces is needed. A submarine has high-pressure air stored in large air banks (tanks) surrounding her hull. These air banks are always kept charged in case of an emergency where the submarine may require them to use the air to literally blow them to the surface. Normally one bank of five is put on service, but if an emergency strikes all five banks are used at the same time. In general, controlling these banks determines how a submarine dives and surfaces. The top of a submarine has large vents that open and close for each ballast tank. To stay surfaced the vents are kept shut and air is trapped in the tank. When a submarine wishes to dive the vents are opened, allowing air to escape and be replaced by water, to surface the vents are closed and high pressure air forces the water out of the bottom of the ballast tank which is always open to the sea. This is much like a glass floating in water upside down with air trapped in it, as long as it is not tipped over, it will remain floating. A submarine works on the same principal. High-pressure air moving at an extreme speed to fill the ballast tanks with air often freezes the piping and valves associated with it. This was ultimately the reason the submarine was lost.

The Skylark's last recorded communications with the Thresher was at 9:17 and it was a garbled message, though some believe it ended with the words "test depth." Some on the Skylark also say they heard through the communications the sound of high-pressure air. Seconds after this communications the distinct sounds of a submarine breaking apart was heard on the UQC (Underwater telephone). The Skylark continued to try to communicate with the Thresher but no further communications occurred. It is known that the submarine sank in approximately 10,000 feet of water with all 129 men lost.

The Skylark was an auxiliary ship with limited communications equipment on board, and was not fitted with the high technology equipment of a war ship. The Skylark actually took

1 hour and 58 minutes to gets its alert message through to the Commander flotilla 2 at New London, Connecticut. Only then was the search and rescue procedure set into motion. What happened to the Thresher was not yet a critical situation according to those in New London. Loss of communications, in fact, was a pretty common occurrence. After a review of all the communications, and not till then, was it determined that the situation did not look good and an immediate concern became apparent.

A 1963 CNN article reported it was a mechanical fitting that let go under immense pressure which caused water to blast through the hull with such force that it may have reached temperatures able to melt metal. In either case it appears the Thresher had a problem more likely related to the fact that she was undergoing tests for new equipment installations and systems, especially the sea water systems and air pressure systems which are tested for proper performance during any shakedown cruise. As with anything new it must be checked to see how beneficial and reliable it will be under real conditions. Obviously, in the case of a submarine, this can prove to be very hazardous, if not fatal, as it was with the Thresher.

In this case study it is obvious that the Thresher had problems and the official hearings by the Joint Committee on Atomic Energy pointed it out. What is not obvious is whether her problems were really not that unusual, compared to other submarines of her era. The submarine had problems which were repaired and received normal overhauls to maintain her equipment. The Thresher was only in commission a little over two years when disaster struck. During any overhaul, a submarine has a mixture of new and experienced personnel assigned to her. Training is constant, but most of the new personnel may never have been at sea before. The Thresher herself had not been to sea for nine months. When all of the possibilities for her disaster are taken into account, could it have been the shock testing done earlier that caused the submarine's failure? Remembering that the Thresher was subjected to more shock testing than any other submarine in United States history has also been considered and often thought about, but no definitive answers have been given. What this case study gives the reader, unfortunately, is many possibilities or reasons for the Thresher's loss.

The reader is invited to contemplate the many different arguments for the Thresher loss and make opinions based on the many factual results given as to the reasons why this disaster happened. It is fully understood that reaching an absolute positive explanation will probably not be forth-coming.

What can be understood however was that the United States Navy would forever change the way submarines would be built, overhauled, and maintained, as lessons learned from the Thresher disaster were far reaching. The biggest and most important change after the disaster was the implementation of the Submarine Sub-Safe System. This system was split into five major categories: piping systems, flooding control and recovery, documentation, pressure hull boundary, and government-furnished material. These systems were now strictly monitored and had to pass specific safety tests and quality control requirements before being adjusted, repaired or replaced. In addition any associated equipment that came close to approaching a sub-safe component also was subject to higher quality control. This included pre-installation, during installation and post installation inspections. This was followed by a series of tests to insure the Sub-Safe System worked according to specifications during actual at sea trials. Only then did everyone sign off on the Sub-Safe System.

The systems, quality control, sea trials, welding, and x-raying of the hull welds had a better commitment for and to safety then ever before. The checking and re-checking of all vital systems would be better watched, looked after, and checked than ever before. Submarine sea trials, after a shipyard overhaul, would also be given more hours of attention to detail than in the past. The Thresher disaster was responsible for making submarines safer in the future.

The one single positive point that might have come out of the Thresher loss was that safety was instantly updated with new requirements and equipment, including the new Sub-Safe System which was a direct development because of the Thresher. It is a fact that the loss of life on the Thresher was a terrible acknowledgement to have to witness. Ironically, however, it probably saved many future submariners' lives and put newer submarines into service. Norman Polmar's book, *The Death of the USS Thresher,* outlines several possible scenarios as to what happened during those final terrible moments. (124-125) What really caused the demise of the Thresher will probably never be known, because the answers lay somewhere with the pieces of the Thresher that lay deeply buried in the silt and mud on the ocean floors. What the official report shows is that inadequate building, maintenance, and testing procedures cost the lives and loss of the Thresher. The one good point is that it finally made the Navy more aware of these problems, and it set into motion a plan to correct and improve on the problems that caused her loss.

Chapter 5

Intelligence Damage

In 1960, the United States implemented SOSUS (Sound Surveillance Systems) which consisted of hundreds of sensitive listening devices sunk in both the Atlantic and Pacific Oceans to monitor transit lanes, and ship and submarine movements. This system was also able to monitor and pick up numerous problems with submarines. Signatures of noise were transmitted back to the United States for immediate investigation and eventually allowed the United States to develop a large library of data signatures. This library was explicit enough to allow the United States to track individual ships and submarines by the noise (signature) they generated. The system was sensitive enough to detect faulty bearings in motors, a chip in a propeller, or noise of any other problem related component. SOSUS also alerted the United States to explosions and was one of the chief tracking devices used to detect Russian submarines that sank due to equipment failure. Ultimately this system detected a number of submarine accidents and losses that occurred. Unfortunately for the United States this system was compromised when John

Walker and his family of spies gave away many top secrets that dealt with the United States submarine force.

John Anthony Walker Jr. was born on July 28, 1937, in Washington, D.C. and was a Warrant Officer and Communications Specialist while in the United States Navy. He also served as a Radioman on two of our ballistic missile submarines the USS Andrew Jackson SSBN 619 and the Simon Bolivar SSBN 641, so he had access to sensitive information the Soviets were more than happy to get their hands on. Walker also fit the profile of the typical person the Soviets would look for in a spy. He was severely in debt (thousands of dollars) and certainly could use the money the Russians would offer him. In fact, as time went on according to Pete Early a former reporter for the "Washington Post" and author of the book *Family of Spies*, Walker solicited his wife and children's help in spying for the Russians because greed and money were something the family had and needed.

Another area that must also be considered is what effects the John Walker spy case had on the other services? Walker's now ex-wife, Barbara, was one of the prime witnesses against him during the trials that followed and provided most of the evidence against the Walker family spy ring. Walker started his spying in December 1967, but was not arrested by the FBI until 18 years later. Walker's ex-wife tipped off the FBI, because she was annoyed about not receiving alimony, which led to his arrest. Walker himself stated that "K-Mart has better security than the Navy did." This comment makes one wonder whether this was sour grapes or something that this country should have been much more concerned about. Though some of Walker's family (Michael, Barbara, and Arthur) was able to cut deals with the government, John Walker however, was found guilty and is now serving a life term in Springfield,

Missouri as prisoner number 22449-037 and is housed at the U.S. Medical Center for Federal Prisoners (MCFP).

It was not until his capture that the United States learned that Russia had penetrated the U.S. Navy submarine codes. These codes were given to Russia by Walker, and his submarine service as a radioman certainly provided him with valuable information that Russia was very happy to receive. This betrayal by Walker may very well have caused the demise of the Scorpion according to Kenneth Sewell, author of the new book *All Hands Down*, because it allowed Russia to determine where and what the Scorpion was doing at the time of its loss. It is known that the Scorpion exploded at 6:44 PM and sank in more than two miles of water, about 400 miles from the Azores. To date, the United States Navy still has not released the full accounting as to the exact cause of the loss of the Scorpion, nor does the United States Navy discuss any Cold War tension if any existed during that time frame.

On the other hand, unlike the missile firing submarines that had two crews, the blue and gold, the fast attack submarines only had one crew, spent more time at sea, and had much less upkeep and repair periods. This brings up still unanswered questions about the Scorpion and Thresher. Could better training and more upkeep periods have prevented these submarines' losses or was intelligence damage to the SOSUS system the key factor? Why is it that supposedly these two submarines still lay on the bottom of the world's oceans? It is known that their wreckage has been found because the United States has several pictures that were able to pinpoint their locations on the ocean floor. It is also known that submarines can be raised, as was the case with the Russian submarine Kursk. It obviously is a question that does not have any simple answers. The condition of the submarines, the state of damage after hitting the bottom, and of course how deep they are all play a role in raising them. What is not clear is whether it is intelligence concerns or the condition that they are in that keeps them from being raised.

Intelligence damage to the SOSUS system certainly must have affected the United States during the Cold War Era, especially since it was not something many other countries knew about and when compromised allowed them to realize that submarine and surface ship movements could be tracked and was no longer a secret. It is also possible that the Soviet Union may have had an upper hand in determining the United States submarine operations during this time period. Whether the damage was a direct cause of the loss of the Scorpion in May 1968 will probably never be known or will certainly not be released to the public, but it will remain a discussion point for years to come.

In January 1968 the North Koreans seized the United States Intelligence Ship Pueblo. One of the items confiscated from that capture was a cryptographic unit capable of deciphering the United States Navy's top secret codes. At the same time, unbeknownst to the United States, was the fact that Walker and his partners were supplying Russia with the codes. (*All Hands Down*) The intelligence codes were compromised and much damage was done. Was the Scorpion loss part of this damage? The codes supplied the information that was needed to decipher American encryption machines, especially the model KL-47 an older model and a new state of the art model the KW-7. Walker even was able to turn over technical and repair manuals on these machines by copying them from the Simon Bolivar's radio shack. These machines were widely used for Naval Communications thus allowing Russia to learn what these communications contained, specific submarine schedules. (Sewell, *All Hands Down*)

Historians Kenneth Sewell and Jerome Preisler's book, *All Hands Down,* describes what they believe happened to the Scorpion. Sewell was an experienced submariner and laid out his beliefs that a torpedo was the cause of the submarine loss. Kenneth Sewell served on the USS Parche, also a fast attack submarine. He ultimately conducted and scrupulously researched the Scorpion's demise and strongly disagrees with the published reports from intelligence gathering and the court of inquiry as to the reasons for Scorpion's loss. The official Court of Inquiry finally released to the general public on October 26, 1993 stated that the loss of the Scorpion was still unknown. Sewell and Preisler state in their book, "In March 1968, a Soviet sub, K-129, mysteriously sank near Hawaii, hundreds of miles from its normal station in the Pacific. Soviet leaders believed that it was mistakenly sunk by a United States submarine." They also state "that from that moment on the Russians planned on revenge and deliberately set a trap by having a large exercise of ships act suspiciously in the Atlantic. They knew it would only be a matter of time before the United States would dispatch a submarine to investigate." Sewell and Preisler's book seems to fall in line with an older book on this subject by Edward Offley, *USS Scorpion-Mystery of the Deep,* who also believes that the Russians were involved with the loss of the Scorpion and all ninety-nine crew-members. These authors all believe that Russia was the cause of the Scorpion's loss. Sewell and Preisler put it this way: "Scorpion's crewmen were killed while defending themselves against a blatant act of war, and we all share in owing them a debt of gratitude." Far too often reports are received that dictate a cover-up. Many lead away from the real truth like the highly placed U.S. intelligence source involved in the United States Navy's official Scorpion investigation, stated in the book *All Hands Down,* "Ninety-nine men aren't worth the destruction of civilization. Other reasons had to be found for the loss." The politics of the Cold War continues to play a very secret role as far as telling the truth is concerned. Except for the occasional book, historian, or critic the Scorpion stays the mystery that the United States government refuses to divulge to the public.

It is safer to say that most of the Scorpion information is old and forgotten, except for those directly involved, like the families who still live in the darkness of what the real truth is and will never really know what caused the Scorpion loss. Intelligence reports seem to dictate from various leaks that more is known than told. In *All Hands Down* one conflict in published reports appears to be validated by one of the pictures of the sail on the Scorpion. In this picture it is easy to see that the sail of the submarine appears to have a section missing that could not have happened without something hitting it. While it is possible that the submarine could have run into something it is unlikely that an entire section of the sail would be missing as the picture shows. It is Sewell's opinion that this is where a torpedo from a Russian sub hit the Scorpion. At the very least there are significant questions which have not been convincingly answered.

Also the arrow in the picture shows that the submarine had an antenna raised, which dictates that she was near the surface at the time of the accident. Intelligence damage caused by Walker's ring and the subsequent knowledge of classified key codes obtained by the Russians will always be open to scrutiny in whether they played any role in the sinking of the Scorpion.

Chapter 6

Environmental Impact

The Cold War brought about the building of submarines at an incredible rate. Many of these submarines are no longer on active duty, which brings up the question, what to do with them now? What must be done with them at the end of their useful life? The environmental dangers possible if not properly disposed of because of their reactors are a main concern, especially since many have not been defueled. The defueling process removes the nuclear fuel and rods from the reactor vessel, which disables the majority of radioactive dangers to the environment. The core must then be cut out and stored in a safe containment area along with the nuclear fuel and rods that are left over. This is where the concern comes in, as a disposal method for the defueled reactor compartments is needed when the cost of continued operation is not justified or when submarines have been decommissioned. After a nuclear-powered sub is no longer needed, it can be placed in protective storage for an extended period (usually waiting for an opening to allow dismantling) followed by being permanently disposed of or recycled, or prepared for permanent disposal or recycling immediately after being decommissioned. This extended period often is not more than one year and the submarines are kept tied up at a pier, usually at the site where they are to be dismantled and with a skeleton crew aboard. The preferred method is land burial of the entire defueled reactor compartment at the Department of Energy Low Level Waste Burial Grounds at Hanford, Washington. A ship can also be placed into an indefinite floating state where the ship is put into storage for a long time without risk to the environment. Every few years each ship would have to be taken out of the water for an inspection and repainting of the hull to insure that the hull does not rust and is checked for continued seaworthiness. However, this protective storage does not provide a permanent solution for disposal of the reactor compartments from these nuclear-powered submarines. This is the state of many of the Russian submarines from the Cold War today of which there are over 100. Thus, this alternative does not provide permanent disposal and requires security and maintenance procedures to be put into effect throughout the entire storage period.

Today the inventory of submarines that need to be dismantled continues to grow. Russia presently has 170 submarines that are sitting at its piers in various states of disrepair because no money has been allocated to continue the dismantling process. This dismantling usually takes place on Russia's Kola Peninsula at one of three ports, Zapadnaya Lista, Vidyayevo, and Severmorsk. These ports are important to Russia's Northern Fleet especially since they are ice-free and handle the majority of its submarines. (Northern Fleet Overview) Because of the backlog and the fact that Zapadnaya Lista is Russia's largest submarine base it handles the largest amount of dismantled submarines. The problem continues to be the fact that the state of its dismantling program has not changed very much in the last 10 years due to Russia's continued lack of funding to the program. This translates into a large concern that at any given time the potential for a nuclear catastrophe in this city of about 30,000 people exists. Most of Russia's submarines from the Cold War Era have two nuclear reactors and this just adds to the problems already faced with the dismantling process.

According to Georgi Kostev, a member of the Committee of Moscow's Critical Technologies and Non-Proliferation:

"The fact remains that the Navy does not have enough money for the conservation of ships withdrawn from active service. Under lack of financing, the Navy lacks funds for maintaining combat ships in proper condition." This means that the Russian Navy is not interested in allocating money in maintaining the decommissioning of ships reliably. Simply put, the Navy is not interested in investing or preserving the natural environment of its country.

This attitude and lack of concern is a nightmare that spells trouble and outlines a disaster waiting to happen in Russia. The military must set priorities. According to Tatko and Robinson, researchers from the Monterey Institute of International studies, Russia has laid up 104 decommissioned submarines with 72 still not defueled. This defueling process includes: removing the submarine from commissioned status, removing the nuclear reactor electrical circuits and extraction of spent nuclear fuel, transporting this spent fuel to a waste depository, containment of all radioactive waste and any necessary associated cleanup after the spent fuel has been transported.

In comparison, the United States has funded the decommissioning of its submarines and is not faced with the same problems as Russia. The United States primarily dismantles its nuclear submarines in one of two places,

Portsmouth Naval Shipyard in Kittery, Maine or the Bremerton Shipyard in Washington State. The United States follows a strict time scale of the decommissioning of its submarines with a range of 20-30 years depending on the class and deployment schedules.

When a United States submarine is decommissioned it is transported to one of these two shipyards and the dismantling process begins. There is very little time between decommissioning and the actual dismantling process, so that the submarines are not permitted to deteriorate. In addition, the submarines' crew stays with them during the changeover period from active duty to dismantling to ensure qualified, properly trained and competent personnel are on the submarine during this critical stage.

While the dismantling of submarines is not a priority in Russia, when it does occur, one of the biggest problems is the fact that the majority of the time it is left to Russian crewmembers who were not fit (drop- outs) for active submarine duty. They have little or no training, and certainly are not competent in the job of dismantling submarines. Most of these submarines are tied up at piers waiting dismantling with no knowledge as to when that will happen. When comparing the two countries, it is easy to understand what the world now faces from their submarine dismantling programs. The differences are enormous and the safety concerns associated with these submarines' dismantling programs will not end anytime in the foreseeable future.

"The United States has a complete system in place to undertake this work and the associated funding to go along with it, while Russia lacks both of these key ingredients," as stated by the Arms Control Association. The United States presently cuts out the reactor compartments of its nuclear submarines and transports them by barge and train cars to its disposal site in the southeast part of Washington, to its Hanford Nuclear Reservation burial lot. The lot is on a plateau about seven miles from the Columbia River. Hanford is 586 square miles of flat desert. Eight burial sites are located on 518 acres. It received its first reactor core in 1986.

Also, according to the association, Russia produced 244 nuclear submarines that account for 52 percent of the total built worldwide. In the past, according to the association, Russia dumped all of the nuclear waste into the sea, in containers not properly checked for proper containment and only recently (within the last several years) started to abide by the London Dumping Convention, of December 29, 1972, which forbids dumping nuclear waste at sea. Because of the lack of funding and the overstock of piled up submarines waiting for dismantling, it is easy to understand the potential for environmental problems. Many of the Russian submarines are tied up together and though not necessarily as dangerous by themselves, their close proximity implies a disaster of unprecedented proportions if just one should have a reactor malfunction.

The Navy, in a 1999 document titled, "U.S. Naval Nuclear Powered Ship Inactivation, Disposal and Recycling," stated, "After burial, direct radiation at the land's surface will be insignificant (i.e. below detectable levels) due to the low contact radiation fields on the package and the shielding effect of the soil cover." The problem however, in an article by Jack Dorsey, a reporter for the Virginian Pilot, called "Nuclear Ships: Millions to Build, and Now Millions to Trash," states that the Hanford site must be constantly monitored for radioactive leakage, and actually has had some leakage into the Columbian River, though not a big concern at this time. This statement continues to show just how hard it is to properly prepare for the dismantling of nuclear submarines and keep the world population safe from radiation fallout not only during the initial burial of the reactor vessels, but also for the hundreds of years that they present a hazard to human life. Russia is finally beginning to follow the United States by burying its reactors at various sites in Russia.

Overall the concern for proper storage, proper monitoring, and proper safeguards will continue to be a worldwide concern for years to come. The biggest concerns however, are not that the technology to safely contain reactor waste isn't possible, but rather will those that must dismantle their submarines do it correctly when no one is watching? This is the question that will always be left behind following the decommissioning of nuclear submarines. Nuclear submarines in all countries that reach the end of their life cycle must be dismantled properly and safely. A plan needs to be put in place to accomplish this when they are being built, not as an afterthought.

Chapter 7

Reflection on Twenty-six Years
Submarine Cold War Experiences

Long before the first submarine ever came into my life, as a young high school student the Navy and submarines made an impression on me. I had on occasion been able to visit several of the old fleet type submarines in New York and Baltimore in 1966 which immediately sparked my interest and paved the way to what would become my future and career. The submarine was very mysterious to me and I wanted to learn more about them. While in my senior year at high school I did a report on the USS Nautilus and spoke to a Navy recruiter who had been in submarines earlier in his career. He told me about the missile submarines and the fast attacks. He also told me that they were the elite and were provided with the best food in the navy. What stuck in my mind however, more than anything else was the continued educational opportunities that he said would be provided to me if I joined the submarine service. He told me that I would be sent to school often to stay abreast with all the new technology. To this day I am still fascinated with the submarine service that has become such a large part of my life, and continue to go on board them with every chance that becomes available.

There was another day that will always remain in my mind. It was a rather cool summer day in Groton, Connecticut and my step son and I were out at Rock Lake at the Submarine Base in Groton, learning how to dive and surface a remote controlled model submarine. On that day we had no problem with diving the model and just like on the real submarines it submerged at my request as the captain of that model. But that was where the fun ended, because the model did not come back up to the surface when it was asked to. Not only did that yellow submarine not come up, here I was a Qualified Diving Officer on a fast attack submarine unable to bring up a simple model submarine. Needless to say the rock lake water was very cold that day. On a more serious note however, this problem intrigued me, because it made me realize what real dangers submariners face every time they dive. Unlike the model where you could go into a lake to retrieve it, a real submarine that is lost never comes home again. It brought back the memories of the Scorpion and Thresher and all of the other lost submarines that are still on patrol today.

My submarine Cold War history began in 1967 when I became a submarine school student at Groton, Connecticut. Submarine school is the initial step that submariners are required to attend before being assigned to any submarine. It is in this school where students learn about the basics of submarine systems, which include such things as learning about hydraulics, high pressure air, freon control, steering and diving systems, reactor controls, radiation, announcing systems, propulsion, qualifications, and the requirements needed to meet the high standards necessary to earn the precious silver or gold dolphins which signify qualifications in submarines. This is a long process that sometimes takes several years to complete because the student must not only learn and have knowledge of these systems, but must also demonstrate that he understands how each system works and ties into the other systems on the submarine. Each system supports another system in much the same way that a clutch supports the operation of a transmission in a car. A submariner must learn every system, and every valve including how to operate them. In any career, which may span many years and many different submarines, this process is started every time a submariner is assigned to a new submarine. In the 26 years of my career I had to re-qualify on each of the six different submarines I served on. This, as I stated earlier, is necessary because there are many different classes of submarines and though the main systems are the same, location and operation are different. Newer submarines continue to expand on technological advancements that are incorporated to improve upon their operation and because of this a submariner must continue to stay abreast of these changes continuously devoting time to their schooling, studying, and training.

In 1968, after graduating basic submarine school, I reported on board the first of my six submarines which was one of the soon to be called "Forty-One for Freedom" submarines, the Theodore Roosevelt SSBN 600 (submarine service ballistic nuclear). This was the third ballistic missile firing submarine built by the United States, and became my home for the first four years of my submarine experience.

Theodore Roosevelt SSBN 600
(Picture courtesy of U.S. Navy)
Specifications in Appendix A

Qualifications started immediately and many hours were spent studying the systems and various components that made up this submarine. There were high pressure air, hydraulic, gas, fuel, battery, electrical, electronic, diesel, reactor, sonar, radar, and many other systems to learn. These systems would all have to be demonstrated to ensure you understood the basics of their operation and before you would be allowed to earn dolphins, the service emblem worn on my uniform to show that I had qualified in submarines.

My time would also be spent learning a new way of life. Day and night were no longer a part of my daily experience but my body soon adjusted to the 18-hour day to which all submariners become accustomed. I was required to set aside a certain amount of time during each 18-hour period to study, eat, and stand watch and sleep. In addition I had to maintain a weekly goal (be interviewed on so many systems per week) so I did not become delinquent, as this would affect my free time if I did not keep up with my studies at the required rate.

My first experience at sea came in February 1969, when, for the first time, I would actually get to experience what had to be done to prepare a submarine for the open sea. The crew was split up into specific areas of responsibilities. Each system had to be checked and readied. (Many systems are shut off while a submarine is tied to a pier.) During this process some components may fail, need calibration, or even need to be replaced. A checklist is gone over for each and every compartment on the submarine. Everything is checked and rechecked, verified and than verified again. Each crew member must sign off on his area of responsibility. Once this is done a second person will recheck what the first person checked to ensure a second margin of safety. This practice is carried out over the entire submarine starting 48 hours prior to getting under way. In addition each compartment will have specific instructions posted in convenient places that outline what must be done for any causality whether in port or at sea. During this learning experience I quickly became overwhelmed, but I also was amazed at how everyone became team members and worked well together.

Finally it was time to get under way (leave port). I was assigned as a helmsmen and plainsmen during the maneuvering watch. This meant that I would be steering and diving the submarine under instruction until I became qualified and proficient in this watch station. A maneuvering watch is the assignment that all crewmembers are on while the submarine prepares to leave port. It puts experienced men at specific stations during this initial time as the submarine gets under way. I was amazed at how many people were involved and required just to get the submarine under way. While the submarine started moving, I was given instructions and watched to ensure that they were followed correctly. The rest of the crew was following similar procedures with confidence, which suggested they had done this before and were well trained. I spent six hours steering whatever course instructions I was told to follow. I also quickly built up a confidence in my new job, and I was pleased to be put in such an important position at the young age of 19. This however, was not unusual as many young crewmembers are still in their teens when they first report aboard a submarine; this has been in practice since World War II.

The Cold War ballistic missile submarines' sole purpose in life is to remain undetected and prepared for firing its missiles at a moments notice if called to do so. It is not to seek out other submarines or gather intelligence, though if the chance occurs it will do so. Most of the time it is assigned a patrolling area to monitor and it stays in this area, approximately 70-75 days, ready to act when called upon. These submarines carried 16 missiles with multiple warheads capable of destroying several cities with each missile. My first two patrols were spent qualifying on the ship's systems, so I was kept very busy.

Submarines are not made for human comfort. The sleeping quarters are barely big enough to sleep in and even turning over is a challenge. Some human comforts are accommodated however. The berthing areas do have stereo systems with selectable stations for listening to music. Showers, unlike aboard the old diesel submarines, are allowed since fresh water is made from sea water instead of just being stored. The food on submarines is the best in the Navy, and rivals being the best in any of the services. Mealtime on a submarine is one of the main past times and is a priority when it comes to maintaining crew morale.

The missile submarines have two crews called the blue and gold. Each crew alternates between spending three months on the submarine and three months in port going to school and getting some rest and relaxation. This patrol schedule has each crew making two patrols a year and spending about 150-160 days at sea. The extra days are used for 25-30 day upkeeps to prepare the submarine for going to sea.

These in-port stays also include repair and maintenance of equipment and crew turnover. One of the advantages to this type of schedule is always being schooled to stay on top of the latest technological advancements, since the submarines are updated with new equipment at this time. Schooling ensures that the crew keeps up with the knowledge needed to operate, troubleshoot, and maintain equipment. In addition the schooling had another big advantage; it helped in being promoted, often earlier than those in other branches of the services. I served on two other missile submarines during my first 13 years. These submarines were the Tecumseh SSBN 628 and the Daniel Boone SSBN 629.

USS Tecumseh SSBN 628
(Picture courtesy of U.S. Navy)

USS Daniel Boone SSBN 629
(Picture courtesy of U.S. Navy)

The three missile firing submarines were all pretty much the same in layout, though they were two different classes of submarines. The Roosevelt was from the first class of missile firing subs called the Washington class, of which there were five built. The Ethan Allan class (five built) was an evolutionary development from the George Washington class which was followed by the Lafayette (nine built), the James Madison (ten built), and the Benjamin Franklin classes (12 built). These classes made up "The Forty-One for Freedom." The Tecumseh and Boone were part of the James Madison class of submarines. The main differences between the classes was the size, 382 feet (Washington class) to 425 feet (Franklin class), and the advancements in technology for sonar, radar and other various electrical/ electronic equipment. There were also many advances in sound quieting.

My upkeep periods were all at Holy Loch, Scotland. During my 13 years on the missile submarines I made many patrols and continued to increase in rank (promotions) and responsibility. After my initial patrols I became a striker in the Interior Communications Electrician field. A striker is a sailor in the Navy who is trying to learn a field or job opportunity without formal schooling for that job. It requires self-motivation, as self-study and hands on training supply all the instruction. When there is an opportunity or opening the Navy allows this option for picking or changing the job field you are in as long as you are willing to supply the effort. This was how I got into a field and career path of my choice. For the 13 years that I served on missile submarines I advanced both in rank and responsibility and soon was in charge of the Interior Communications division. The job included maintaining all of the submarines electrical communication servo control systems, instrumentation, batteries, phones, gyroscopes, atmosphere monitoring equipment, air and hydraulic indicators, and various other communication systems. It also required training and maintaining the morale of all of my division personnel.

My missile submarine days ended in my 14th year because once I became a Chief Petty Officer I was re-assigned to my first attack submarine. The attack submarine is quite different from its missile submarine counterparts because its mission is to seek out and destroy enemy ships and submarines in times of war, patrol our coastlines and gather intelligence during peacetime. The fourth submarine that I was assigned to was the fast attack submarine, the USS Whale SSN 638 (Submarine Service Nuclear). In my opinion, the fast attack submarines are much more challenging and interesting.

The USS Whale was a Sturgeon class submarine, and the second of this class to be built.

USS Whale SSN 638
(Picture courtesy of U.S. Navy)
Specifications in Appendix B

The Sturgeon class of submarines was smaller and faster than the huge missile submarines. They also had the capabilities of breaking through the ice at the North Pole. The fairwater diving planes that were mounted on the submarine's sail (where the antennas were protected) could be turned straight up and down; this helped to break through the ice while surfacing at the North Pole. During my time on two of the Sturgeon class submarines I qualified as a Chief of the Watch and Diving Officer of the Watch, which are in the control room of the submarines.

The Chief of the Watch monitors the conditions of all the vital components and instrumentation on a ballast control panel and makes changes in the ship's trim as directed by the Diving Officer of the Watch. He also monitors the ships hydraulic plants, air systems, hull openings, water levels in variable trim tanks, antennas, phone systems, and several other vital ships systems. Basically he is the controlling monitor for what is happening in the submarine at all times while at sea, and most communications are directed through him.

The fifth submarine to which I was attached was the USS Trepang SSN 674; also a Sturgeon class submarine and was by far the best submarine on which I served. The Trepang, in 1985, made a northern run and broke through the ice several times at the North Pole that spring. When breaking through the ice the submarine must hover (maintain absolute underwater level control) and touch the ice with the top of its sail. It then will use a computer to control the hovering system to push the submarine through the ice using high pressure air forced into a large set of tanks called damage control tanks. This offsets the submarine's neutral buoyancy making it more positive and the sub forces itself through the ice, making this a very delicate maneuver.

The northern run in 1985 was the most interesting experience during my entire 26 years on submarines. The Trepang had a crew that worked hard together and accomplished many firsts during her career. Intelligence gathering, mapping and charting areas in the Atlantic were also among her accomplishments that are seldom spoken of, but equally important. The Trepang, seen here with me in 1985, demonstrates her surfacing at the North Pole.

Trepang and Mark Pater Noster, May 1985, North Pole

Many days on the submarine were spent practicing and drilling to improve our efficiency. Calibration of all the ship's monitors was also very important during this evolution. Much work was done to ensure that every piece of equipment was operating at peak performance, as top conditions are crucial during this time. Watch standing, though always vigilant, was even more so on this run, and more dangers were present once we were in the ice-covered areas of the North Pole. Sonar and radar systems had extra watch standers, as did many other compartments in the submarine. The entire efficiency of crew and sub was at their best operating condition in years. Icebergs and underwater ice were particularly worrisome to the Trepang. The submarine preformed this type of operation to demonstrate the United States' ability to operate not only under the ice, but also the ability to break through it if necessary. During the Cold War Russia and the United States both played this game of hide and seek using the ice as a way to hide their efforts and missions from one another and at the same time gather all the intelligence on each other that was possible.

My sixth and final submarine was the USS Glenard Lipscomb SSN 685.

USS Glenard Lipscomb SSN 685
(Picture courtesy of U.S. Navy)
Specifications in Appendix C

She was a one of a kind fast attack submarine with electric drive. The Lipscomb was unique in more ways than one. She was only the second submarine to tryout a propulsion system with turbo electric drive. She also was the only submarine in the United States Navy named after a congressman, Congressman Glenard P. Lipscomb (1915–1970). The electric drive system, however, did not work out due to many disadvantages, one of which was that the equipment used for propulsion was much heavier, making the submarine much slower. The electric drive caused a considerable problem throughout the submarine with carbon dust (highly conductible particles) dispersed from the electric motors and played havoc with the electrical equipment on board. These disadvantages led to the decision not to use this design on future submarines. She was, except for the engine room, very much like the Sturgeon class submarines, and combat ready. She was also one of the quietest submarines in the Navy, which made her very hard to detect.

My time on the Lipscomb was spent standing watch as a diving officer. The diving officer is responsible for the control of the safe diving and surfacing of the submarine and its safe navigation and depth control while maneuvering underwater. It is a very interesting job as one must learn about such things as how much water needs to be moved in and out of the submarine to maintain negative, positive, or neutral buoyancy. He also must learn about the effects of water temperature and salt water content (salinity) of the ocean as they both can change the weight of the submarine while it moves through the water. All weight on the submarine; both crew and equipment must be accounted for at all times.

I spent many hours studying to be certified (qualified) as a diving officer. After several exams and instruction under the control of other qualified diving officers, I had to pass both an oral and written exam from the Captain of the submarine. The diving officer must not only know about the mentioned items, but also what must be done in any emergency. My exams, and the captain's interview, always centered on emergency procedures that included questions on flooding, fire, reactor scrams, loss of propulsion, and loss of electrical systems, which all call for different procedures to be followed at a moments notice because time can mean life or death in an emergency.

In my entire 26 years on the Navy's submarines I never stopped learning or studying. There were the sacrifices of not being home with my family while spending many weeks at sea, but the rewards from being part of this elite service more than made up for the time away. The education received was paid for by the Navy and the friendships and memories will last forever. Today I look back on the Cold War Patrols with a sense of accomplishment. The education (classroom) and hands on schooling from the Navy in many ways out performed the normal traditional college education that I received from various colleges. This is not to say that they were not valuable, but I found the hands on training to be a better method of instruction that could be applied immediately to my job. Traditional education today, on many occasions, does not allow for working in the field you study for a variety of reasons. The fact that you can apply what you learn almost immediately certainly is a benefit in my eyes. It made the schooling feel more useful when I could apply what I learned in the jobs that I performed.

This memoir of the submarines that I served on and my continued interest in the Submarine Service today demonstrates a feeling that does not lessen with age, nor do the strong feelings of accomplishment which are aroused from being part of the Submarine Service.

Chapter 8

Conclusion

This book has covered many areas of the submarine service that I believe have been of limited knowledge to the public because of the scope of the mission of submarines in general. Some areas have been covered in various research, but it is my belief that never before have these areas all been brought under one roof. The fast attack submarine and the missile submarines which I was lucky enough to have served on have allowed me to present a different and unique experience to the reader. I have also been able to provide a timeline to the loss of the USS Thresher and arguments that give the reader an insight to what may have caused her demise. In addition and perhaps more of a concern, and my biggest contribution, is to open up the eyes of the reader to how just one family of spies can cause so much damage. Finally the reader is made aware of the environmental and possible health related problems that we could all face because of the lack of proper and careful consideration for the retirement of submarines when they are no longer needed or economically safe to maintain.

The Submarine Service played a vital role during the Cold War. The fact that its mission during the Cold War years was constantly evolving does not diminish the role that the submarine service played in keeping our country free from aggression. As this book points out there were many areas and various roles played depending on the type and classes of submarines. The missile submarines had a significant and different role than the fast attack submarines; however both were equally important in the way their missions were conducted.

Significant accomplishments were also seen during this time period. The most famous of these was the USS Nautilus SSN 571 with its historical message. On the morning of January 17, 1955, at 11 AM Nautilus' first Commanding Officer, Commander Eugene P. Wilkinson ordered all lines cast off and the memorable and historic message, "Underway on Nuclear Power" was sent, making the Nautilus the first nuclear powered submarine. Over the next several years, Nautilus shattered all submerged speed and distance records. This submarine's accomplishments however would be just the beginning. Soon after the Nautilus was commissioned in 1959 the USS George Washington SSBN 598 became the first nuclear powered ballistic missile submarine and our first of the now historic "Forty-One for Freedom" submarines. The submarine service played a huge role as a deterrent to aggression for the United States during the entire Cold War Era.

Unfortunately, during the 1960's we also saw our share of accidents. Two of our nuclear submarines were lost at sea and still remain a mystery as to how they were lost. Though my book points out several explanations and historians argue the merits of various opinions, no clear cut answer has surfaced or been proven, nor have any final results as to their loss been published. What is known and been proven is the fact that the submarine service benefited from these two losses. Submarines today are safer because of the many sub-safe improvements that have been made.

The Scorpion and Thresher tragedies brought about the Sub Safe System. A Sub Safe Envelope was put into effect to insure a safety boundary was adhered to for all external sea valves and high pressure air emergency systems on submarines since the accidents. The United States has not lost another submarine since the Scorpion went down, which is proof that the Sub Safe System that was put into place works extremely well.

The question still remains as to whether the Scorpion causality was because of Russia, with the help of John Walker, or an accident? This question is still being argued today and unless Russia divulges this information, or whether we ever try to raise her, is information that will most likely continue to go unanswered. The guise of secret information will continue to be used as a way to not divulge the truth to the public, with very little that the public can do about it.

Just as important to the many successes during the Cold War are the damaging intelligence leaks that were also a part of this era. The Sosus system and the John Walker case caused a great amount of damage to the security of this country, and once again to what extent will probably never be completely known. This too is part of the unanswered question whether Walker directly or indirectly caused the loss of the Scorpion? It is known that valuable keys codes were turned over to the Russians that would have led them to decipher critical information and give them an advantage should hostile plans been prepared against the United States.

So many questions remain today that continue to go unanswered by our government. Russia and the United States during the Cold war years maintained a constant surveillance over each other. Submarines continued to be built at alarming rates with each new class bettering the preceding one. Nuclear power brought about unlimited submerged times and harder detection for both countries. During the last several decades of the Cold War the destructive power of both sides, because of the submarines, was a bigger concern than ever before. Yet it also was the cause for both the United States and Russia to seek a pause and keep a deterrent attitude towards each other. This attitude kept both sides from initiating World War III.

Finally, the environmental impact due to the amount of nuclear submarines that were built has become a monumental problem to the world. Countries building these war machines gave little thought with what to do with them when their useful life was over. Russia is now just beginning to show signs and more concern about the cleanup and safe decommissioning of its nuclear submarine fleet. Outside pressure and financial help obviously is at the root of this cooperation. The United States, though in better condition with their decommissioning programs, still has a lot to be concerned with as well. The one good aspect about this however, is the fact that it is being looked at and acknowledged.

The submarine with its men, pride, threats and disasters will go down in the history books as one of the reasons mankind continues to survive. Accomplishments will continue and new rules will apply in the future. Hopefully the future will be as successful in keeping the peace as the Cold War Era. Even with the difficulties, losses, and distrust, Russia and the United States, as well as other new countries now joining the submarine powers, must continue to look ahead to maintain the peace. The submarine service can be used to maintain peace, or it can be used as a terrible destructive force of war. Whether success will be measured by peace or war the submarine service, as in the past, will play a major role in determining human survival now and in the future years to come.

Bibliography

The American Heritage Student Dictionary, Houghton Mifflin Company Oct. 2004

Blum, Howard. *I Pledge Allegiance: The True Story of the Walkers: An American Spy Family*, Simon & Schuster Books. 1987

Bowermaster, Jon. *The Last Front of the Cold War*, Atlantic Monthly (10727825), Nov 1993, Vol. 272 Issue 5, p36-45, 6p, 2c; (AN 9404280166). EBSCOhost. Excelsior College, Albany, NY. 13 Dec 2007
http://search.ebscohost.com/login.aspx?direct=true&db=aph&AN=9404280166&site=ehost-+live

Borisov, Sergei. *Another Russian Nuclear Submarine Sinks*, Transitions Online, 12141615, 9/15/2003

Christley, Jim. *US Nuclear Submarines: The Fast Attack*, Osprey publishing, LTD. 2007

Couhat, Jean Labayle. *Combat Fleets of the World*, Naval Institute Press, Annapolis, Maryland 1982

Dorsey, Jack. *Nuclear Ships: Millions to Build, and Now Millions to Trash*, The Virginian-Pilot, 2007

Dunmore, Spencer and Ballard, Robert. *Lost Subs*, De Capo Press, Madison Press, Ontario, Canada 2002

Early, Pete. *Family of Spies: Inside the John Walker Spy Ring*, Bantam Books. 1981

Flynn, Ramsey. *Cry from the Deep*, HarperCollins Publishing, New York, N.Y. 2004

Gaddis, John Lewis. *The Cold War: A New History,* Penguin Press, Inc. 2007

Hutchinson, Robert. *Submarines War Beneath the Waves,* Jane's, Smithsonian, Harper Collins Publications. 2006

Herald, Christian. *The U.S. Navy Conquers "Holy Loch"*, Readers Digest New York, N.Y. July 1963.

Johnson, Stephen. *Silent Steel: The Mysterious Death of the Attack Sub USS Scorpion*, Wiley & Sons Inc. Hoboken, NJ 2006

Kellerman, Stephanie. *How a Nuclear Reactor Works Diagram*, Nelson Institute of Marine Research, Santa Barbara, California. NIMR 2001 http://www.nimr.org/react.html

Kneece, Jack. *Family Treason: The Walker Spy Case*, Paperjacks. 1988

Messersmith, Andrene. *The American Years*, Argyll Publishing, Scotland, July 2003

Kostev, Georgi. *Nuclear Safety Challenges in the Operation and Dismantlement of Russian Nuclear Submarines*, Moscow: Committee for Critical Technologies and Non-Proliferation, 1997

Moltz, James. *Perspectives on International Cooperation in the Dismantlement of Nuclear Submarines*, Moscow, Russia 11 December 1998

Offley, Edward. *The USS Scorpion-Mystery of the Deep*, Seattle Post Intelligencer. 1998 pages 1-19

O'Connor, John. TV View. *American Spies in Pursuit of the American Dream*, New York Times. Feb 1990

Polmar, Norman and Moore Kenneth. *Cold War Submarines*, Potomac Books Inc. Washington, D.C. 2004

Polmar, Norman. *The Death of the USS Thresher*, The Lyons Press, Guilford, Ct. 2004 (124-125)

Sontag, Sherry and Christopher Drew. *Blind Mans Bluff: The Untold Story of American Submarine Espionage*, New York Public Affairs, 1998

Tatko, Jill and Robinson, Tamara, *Northern Fleet Overview*, Monterey Institute of International Studies, 1999

Time Magazine. *Belated Concern* Time Inc. Nov.1985

Trepang SSN 674 and Pater Noster, Mark. *North Pole Photograph* May 1985

Robinson, Tamara and Moltz, James *Dismantling Russia's Nuclear Subs*, Arms Control Association. Washington, DC 2008

Smith, T. J. *Undersea Stealth, Submarining in the Cold War*, PublishAmerica. Baltimore Md. 2004
www.publishamerica.com

Sewell, Kenneth and Preisler, Jerome. *All Hands Down* Simon & Schuster, New York City, N.Y., 2008

Sviatov, George, *The Kursk's Loss Offers Lessons*, U.S. Naval Institute Proceedings, Jun 2003, Vol. 129, Issue 6

Time in Partnership with CNN. *Farther than She was Meant to Go.*
http://www.time.com/time/magazine/article/0,9171,830106,00.html. 2007 (1-2)

Trippett, Frank. *Operation Damage Control*, Time in Partnership with CNN - Monday, Jun. 24, 1985

U.S. Department of State. *Environmental Security Threat Report,* October 2001

Appendix A

Theodore Roosevelt SSBN 600
Specifications

Displacement:	5946 tons surfaced
	6700 tons submerged
Length:	116 meters (382 feet)
Beam:	10 meters (33 feet)
Draught:	8.8 meters (29 feet)
Propulsion:	S5W United States Naval reactor
Speed:	20 knots (37 km/h) surfaced
	25 knots (46 km/h) submerged
Test depth:	700 feet (210 m)
Complement:	139 officers and men
Armament:	16 Polaris missiles
	6 × 21 in (533 mm) torpedo tubes

Appendix B

USS Whale SSN 638
Specifications

Keel Laid:	May 27, 1964
Launched:	October 14, 1966
Commissioned:	October 12, 1968
Decommissioned:	June 25, 1996
Builder:	General Dynamics Corp., Quincy, MA
Propulsion system:	One S5W2 nuclear reactor
Propellers:	One
Length:	292 feet (89 meters)
Beam:	31.7 feet (9.65 meters)
Draft:	29.2 feet (8.9 meters)
Displacement: Surfaced:	Approx. 4,250 tons
Displacement: Submerged:	Approx. 4,700 tons
Speed: Surfaced:	Approx. 15 knots
Speed: Submerged:	Approx. 30 knots
Armament:	Four 533 mm torpedo tubes for *Mk-48* torpedoes, *Harpoon*, *Tomahawk*, and SUBROC missiles, ability to lay mines

Appendix C

USS Glenard Lipscomb SSN 685
Specifications

Awarded:	16 December 1968
Laid down:	5 June 1971
Launched:	4 August 1973
Commissioned:	21 December 1974
Fate:	Submarine recycling
Stricken:	11 July 1990
Displacement:	5813 tons surfaced, 6480 tons submerged
Length:	365 ft (111 m)
Beam:	32 ft (9.8 m)
Draft:	
Power plant:	S5W reactor
Speed:	18 knots (33 km/h) surfaced, 23 knots (43 km/h) submerged
Depth:	1300 ft (400 m)
Complement:	12 officers, 109 enlisted
Armament:	4 × 21 inch (533 mm) torpedo tubes

Appendix D

Abbreviations and Definitions

ACA	Arms Control Association
Air	High pressure 3,000-4,500lbs
AMC	Automatic Maneuvering System
Ballast Tank	Large tanks used to maintain a submarine's buoyancy by releasing or trapping air inside of them
CDR	Commander
CO	Commanding officer
Crush Depth	The depth a submarine implodes (is crushed)
Emergency Blow	High pressure air blown into the ballast tanks to surface a submarine
Forty One for Freedom	41 ballistic submarines built during the Cold War by the United States
HQ	Headquarters
JCOAE	Joint Committee on Atomic Energy
LCDR	Lieutenant Commander
LDC	London Dumping Convention
MC System	Announcing system
SSBN	Submarine Service Ballistic Nuclear (Boomer)
SSN	Submarine Service Nuclear (Fast Attack)

Test Depth	The deepest depth a submarine can go safely
UQC	Underwater telephone
USN	United States Navy
Variable Ballast Tank	Tanks that allow water to be transferred between them to trim a submarine during an emergency
XO	Executive officer